The Root of Reality

The Cosmos Point

Stephen Blaha Ph. D.
Blaha Research

MMXXV

.

ISBN: 979-8-9986071-3-4

Rev. 00/00/01 July 24, 2025

To Margaret

Some Other Books by Stephen Blaha

SuperCivilizations: Civilizations as Superorganisms (McMann-Fisher Publishing, Auburn, NH, 2010)

The Exact QED Calculation of the Fine Structure Constant Implies ALL 4D Universes have the Same Physics/Life Prospects (Pingree Hill Publishing, Auburn, NH, 2019).

Passing Through Nature to Eternity ProtoCosmos, HyperCosmos, Unified SuperStandard Theory (Pingree Hill Publishing, Auburn, NH, 2022).

HyperCosmos Fractionation and Fundamental Reference Frame Based Unification: Particle Inner Space Basis of Parton and Dual Resonance Models (Pingree Hill Publishing, Auburn, NH, 2022).

The Cosmic Panorama: ProtoCosmos, HyperCosmos, Unified SuperStandard Theory (UST) Derivation (Pingree Hill Publishing, Auburn, NH, 2022).

God and and Cosmos Theory (Pingree Hill Publishing, Auburn, NH, 2023).

Newton's Apple is Now The Fermion (Pingree Hill Publishing, Auburn, NH, 2023).

Cosmos Theory: The Sub-Particle Gambol Model (Pingree Hill Publishing, Auburn, NH, 2023).

Cosmos-Universe-Particle-Gambol Theory (Pingree Hill Publishing, Auburn, NH, 2024).

Fractal Cosmos Curve: Tensor-based Cosmos Theory (Pingree Hill Publishing, Auburn, NH, 2024).

The Eternal Form of Cosmos Theory Third Edition (Pingree Hill Publishing, Auburn, NH, 2024).

Fundamental Constants of Cosmos Theory and The Standard Model (Pingree Hill Publishing, Auburn, NH, 2024).

Geometric Cosmos Geometric Universe (Pingree Hill Publishing, Auburn, NH, 2024).

Structure and Dynamics of Cosmos Theory and the Unified SuperStandard Theory (Pingree Hill Publishing, Auburn, NH, 2024).

Black Holes, White Holes, and Superluminal Starship (Pingree Hill Publishing, Auburn, NH, 2025).

The Sequence Theory Foundation of Cosmos Theory and Gambol Theory (Pingree Hill Publishing, Auburn, NH, 2025).

BEYOND TIME AND SPACE Fermion Strata & Consistency Conditions for Sequences, Cosmos Spaces and Gambols (Pingree Hill Publishing, Auburn, NH, 2025).

The C o s m o s as a Physics Organism (Blaha Research, Auburn, NH, 2025)

Available on Amazon.com, bn.com, Amazon.co.uk and other international web sites as well as at better bookstores.

CONTENTS

FIGURES and TABLES

Introduction

The author has proposed that the Cosmos (remembering that there is no time before the Cosmos) begins as a zero dimension point, grows to a one dimension line segment, then grows into a (fractal) two dimension unit square, and then grows into a two dimension space governed by a Master Equation.

The Master Equation was shown to embody the initial mathematical point, the generation of a two dimension piecewise linear unit length, the generation of the fractal Cosmos Curve, the generation of the Cosmos spaces and their dimension arrays, the unification spaces sequence, the coupling constants sequence, and the set of fermion mass sequences.

This book shows the above named parts join to form a unity – a root – that solidifies the direct relation of the parts. It also shows that the Master Equation, which is based on a scaling potential, has an equivalent form with a harmonic oscillator-like potential. The scaling form of the Master Equation gives a sequential spectrum of states that are powers of a constant. The harmonic oscillator form of the Master Equation gives a linear equally spaced, spectrum of states. Together they are consistent and provide the two basic forms of sequences.

The generality of the scaling Master Equation solution is supported by its solution also being equivalent to a Free Field solution for small values of the radial parameter.

The Master Equation is explored further to show the connection of Master Equation solutions to the construction of the two dimension Cosmos (Hilbert) fractal curve as a sequence of piecewise linear line segments. Thus there is a direct relation between the fractal curve and the Master Equation.

The fractal curve has a locality property that places points "near" to each other. Artificial Intelligence 1.58-bit Large Language Models (LLM) also have a locality property that enables efficient processing of data. Remarkably one of the sequences implied by the Master Equation has an approximate 1.58 growth factor. This combination of features raises the possibility that the Cosmos may be an LLM for Physical Reality phenomena – a Cosmic simulation – a possibility that has been suggested in an earlier book.

This book suggests a unity at the base of the Cosmos that commends its attention. It provides a fundamental description of the Cosmos that answers the questions: "Why a Cosmos?" "Why the form of its features?" Physics imitates Reality. It does not replace Reality. The book is an important step to a Physical view of the Cosmos as a simulation.

1. A Master Equation for the Cosmos

Why does the Master Equation have its form?

The Master Equation has a series of features that give it a completeness and wide applicability in Physics Theory. It is a differential equation with the features:

`1. A second order Schrödinger-like quantum differential term.
(Space-time and Relativity do not exist in the two dimension space.)

2. A scaling $1/r^2$ potential that prevents the appearance of a mass scale.

3. An energy term that is quantized.

4. The avoidance of other terms. A most minimal differential equation.

5. A corresponding harmonic oscillator-like formulation that implies a form of stability seen in classical harmonic oscillator studies. In classical Physics many forces are approximated by harmonic oscillator behavior for small perturbations.

6.The generality of the scaling Master Equation solution is supported by its solution also being equivalent to a Free Field solution for small values of the radial parameter.

1.1 Continuity of Cosmos Theory Progression

In previous books we began to closely relate the Master Equation with the Cosmos fractal curve. We showed the solutions of the Master Equation have zeros that coincide with the lengths of piece-wise linear line segments. We showed the solution zeros are equal to eigenvalues of the Master Equation. In this book we study these points in more detail. We conclude that there is an intimate relation between the fractal origin of the two dimension space and the Master Equation which governs it. Thus a unity exists in the basis of Cosmos Theory.

1.2 Harmonic Oscillator-Like Equivalent

In this book we also show the Master Equation, based on scaling, is equivalent to a harmonic oscillator-like equation. They have similar solutions at small distances. They have consistent energy eigenvalues. Thus they are equivalent.

Since harmonic oscillation is known to be stable against small perturbations – additional forces may be approximated as harmonic oscillator terms – we can view the Master Equation as similarly stable. Its solutions and their eigenvalues are stable. Thus the harmonic oscillator form of the Master Equation supports the stability of the Cosmos.

1.3 1.58-Bit Large Language Model (LLM) and Cosmos Simulation

The set of sequences of Cosmos Theory has a sequence with a 1.58 multiplier. This multiplier is numerically the same as a ternary bit value in 1.58-bit Large Language Models (LLM). LLM have a locality feature and a data compression feature that supports Artificial Intelligence applications including simulations.

The fractal curve of Cosmos Theory also has a locality feature. This sort of Locality has led to the use of fractal curves for AI applications such as IP address mapping, robbery, location spacing data, data warehouses, dithering, and other AI applications.

Thus we see the possibility that Cosmos Theory may be a simulation along the lines of AI.

2. The Scaling Master Equation

The Master Equation for Cosmos Theory that we previously developed[1] was based on a scaling $1/r^2$ potential. It led to our Sequence Theory. The sequences consisted of powers of a constant. This chapter describes this Master Equation and its implications.

In chapter 3 we generate an alternate form of the Master Equation, which describes a harmonic oscillator. This form of the Master Equation generates sequences consisting of equally spaced eigenvalues, which are consistent with the scaling solution eigenvalues, although it supports linear sequences. Together these forms of sequences constitute the two basic forms of sequences.

2.1 Justification for $1/r^2$ Potential Master Equation

There are several reasons for selecting a $1/r^2$ potential equation. First we choose a second order differential equation similar to the two dimension Schrödinger equation. A $1/r^2$ potential allows us to avoid the introduction of a mass scale. Thus it is the most reasonable initial choice. Later more reasons will appear for choosing the form of the Master Equation. The scaling $1/r^2$ potential gives

1. No mass scale dependence.
2. A spectrum of powers of a multiplier times an overall constant.
3. A constant that can be scaled out of the Master Equation by scaling the radial coordinate.
4. The capability to scale any energy eigenvalue to one.
5. The generality of the scaling $1/r^2$ Master Equation solution is supported by its solution also being equivalent to a Free Field solution for small values of the radial parameter. See chapter 3.

2.2 The Scaling Form of the Master Equation and Cosmos Theory

We specify it with a simple Schrödinger-like equation in coordinates x and y with a λ/r^2 scaling potential. The scaling potential provides the simplest guarantee of an energy spectrum that is a sequence of powers of a constant.[2]

$$[(\partial^2/\partial x^2 + \partial^2/\partial y^2)/m + \lambda/m^2r^2 + C)\psi = 0 \qquad (2.1)$$

where λ is a coupling constant and $r = (x^2 + y^2)^{1/2}$. In polar coordinates

$$x = r\cos(\varphi)$$

[1] Blaha (2025c).

[2] This equation may be viewed within the context of conformal symmetry due to its local scale invariant potential.

$$y = r\sin(\varphi)$$

we find

$$(\partial^2/\partial r^2 + 1/r\,\partial/\partial r + \partial^2/\partial\varphi^2 + \lambda/mr^2 + mC)\psi = 0 \qquad (2.1a)$$
$$(\partial^2/\partial\varphi^2)\psi = 0$$

and the solution

$$\psi = \psi(r)\Phi(\varphi) \qquad (2.1b)$$

where

$$\Phi(\varphi) = \exp(ik\varphi) \qquad (2.1c)$$

The quantum number k must be a whole integer to avoid irregularities in $\psi(r)$. We set $\lambda = k^2 + n_\varphi^2$. Substituting and separating in eq. 2.1a we find

$$(\partial^2/\partial r^2 + 1/r\,\partial/\partial r + (\lambda - k^2)/mr^2 + mC)\,\psi(r) = 0$$
$$(\partial^2/\partial r^2 + 1/r\,\partial/\partial r + \lambda_{n\varphi}r^2 + mC)\,\psi(r) = 0 \qquad (2.1d)$$

where

$$\lambda_{n\varphi} = n_\varphi^2/m \qquad (2.1e)$$

In chapter 3 we derive an alternate quantization of a harmonic oscillator form of the Master Equation.

2.3 Bohr View of Sequence Equation[3]

Bohr developed an atomic model which successfully initiated quantum theory. We use it to illustrate the physical origin of the angular momentum quantization in n_φ. We now adapt that model to the case of eq. 2.1 transformed to a classical two dimension equivalent with attractive potential λ/r^2. Assuming the existence of stationary, discrete states in the 2-dimension model "atom" of eq. 2.1 with an atomic-like nucleus we assume an "electron" mass m, velocity v and radius r. The centripetal force equals the force associated with the potential:

$$mv^2/r = 2\lambda/r^3 \qquad (2.2)$$

The kinetic energy is

$$T = \tfrac{1}{2}mv^2 = \tfrac{1}{2}\lambda/r^2 \qquad (2.3)$$

We assume the stationary states have a quantized angular momentum

$$p_\varphi = mvr = n_\varphi\hbar \qquad (2.4)$$

implying

$$v = n_\varphi\hbar/mr \qquad (2.5)$$

Eq. 2.4 then implies the quantization of λ with the quantum number n_φ

[3] Abstracted from Blaha (2025b)

$$\lambda_{\text{Bohr } n_\varphi} = n_\varphi^{\,2}\hbar^2/m = n_\varphi^{\,2}/m \tag{2.6}$$

after setting $\hbar = 1$.

This semiclassical result allows the Bohr model radius *not* to be quantized but gives a quantization consistency condition on the coupling constant.

While m appears as a mass factor we may view it as a dimensionless constant since λ is a dimensionless constant.

The quantization of the coupling constant λ reflects the $1/r^2$ potential. This feature is not present in the hydrogen Schrödinger and Dirac equations due to their 1/r potential.

2.4 Eigenvalue Spectrum

The eigenvalue spectrum due to eq. 2.1d is

$$C_n = \text{const}\ \exp(-n\pi/(\lambda_{n_\varphi} - ¼)^{½}) \tag{2.7}$$

The constant n can be positive or negative. We choose n to be negative, setting n = -n based on Case's paper[4] where n and n_φ are independent quantum numbers. The quantization of the C_n eigenvalue spectrum is based on a requirement of orthogonality of solutions (an observation due to von Neumann).

Thus we initially have an energy series in eq. 2.7. We generalize this series to a double series in n and an angular momentum quantum number n_φ.

$$C_{n_\varphi n} = c_{n_\varphi} \exp(n\pi/(\lambda_{n_\varphi} - ¼)^{½}) \tag{2.8}$$

where n is the radial quantum number and n_φ is the angular momentum quantum number. The growth factor for each power series sequence is

$$g_{n_\varphi} = \exp(\pi/(\lambda_{n_\varphi} - ¼)^{½}) \tag{2.9}$$

The radial quantum number n parameterizes each sequence of "energy" values. The angular momentum quantum number n_φ, which we give in the next section, specifies the sequence and its constant factor c_{n_φ}. The index n_φ determines the coupling constant $\lambda = \lambda_{n_\varphi}$ in eq. 2.1, for each specific sequence. See Fig. 2.1.

2.5 The Solutions of the Scaling Master Equation

We define a function w(r) to make the equation Schrödinger-like

$$\psi(r) = r^{-½}\, w(r) \tag{2.10}$$

and obtain

$$(\partial^2/\partial r^2 + \lambda_{n_\varphi}/mr^2 + mC)w(r) = 0 \tag{2.11}$$

where

[4] K. M. Case, Phys. Rev. **80**, 797 (1950).

$$\lambda_{n_\varphi} = n_\varphi^2/m \quad (2.12)$$

and n_φ is a positive or negative integer. We now consider the solutions of eq. 2.11 for small r. The complete solution is a sum of Bessel functions.[5]

The solution for w(r) for small r is

$$w(r) \approx Br^{1/2}\cos((\lambda_{n_\varphi} - 1/4)^{1/2}\ln r + b) \quad (2.13)$$

implying

$$\psi(r) \approx B\cos((\lambda_{n_\varphi} - 1/4)^{1/2}\ln r + b) \quad (2.14)$$

where B and b are constants.

2.6 Zeros of the Master Equation Solution

The oscillating cosines above give us the set of zeros of the solution:

$$(\lambda_{n_\varphi} - 1/4)^{1/2}\ln r + b = \pm\pi/2, \pm 3\pi/2, \pm 5\pi\ 2, \pm 7\pi/2m\ ,,, \quad (2.15)$$
$$= (n + 1/2)\,\pi$$

where n is a positive or negative integer. Solving eq. 2.15 for r we find zeros at

$$r_n = \exp(((n + 1/2)\,\pi - b)/(\lambda_{n_\varphi} - 1/4)^{1/2}) \quad (2.16)$$

If we set

$$b = \pi/2$$

and let $n \to -n$, then

$$r_n = \exp(n\pi/(\lambda_{n_\varphi} - 1/4)^{1/2}) = C_{n,n_\varphi} \quad (2.17)$$

where

$$C_{n_\varphi n} = \text{const}\ \exp(n\pi/(\lambda_{n_\varphi} - 1/4)^{1/2}) \quad (2.18)$$

is the nth energy eigenvalue for const = 1. Using

$$g_{n_\varphi} = \exp(\pi/(\lambda_{n_\varphi} - 1/4)^{1/2}) \quad (2.19)$$

and setting the "const" = 1 in eq, 2.18 and setting $g_{n_\varphi=4} = 2$ in Fig. 2.1 below we find the n_φ=4 sequence gives

$$r_n = 2^n = C_{n,n_\varphi=4} \quad (2.20)$$

2.7 Match of Zeros With Energy Eigenvalues $C_{n_\varphi n}$

Each r_n value in eq. 2.17 is thus a Cosmos space dimension array column length. Thus the zeros of the Master Equation solution map to Cosmos spaces dimension arrays. For positive n we obtain the HyperCosmos spaces. For negative n we find the zeros match the Limos spaces' dimension arrays. In chapter 4 we will see that the

[5] Since we can scale the r coordinate due to the scale covariant $1/r^2$ potential we can take the small r solution to be extended indefinitely.

scaling Master Equation has an intimate relation to the construction of the two dimension space through a fractal process from a point.

2.8 Range of Solution's r Values

We changed the sign of n in eq. 2.8 to obtain the usual form of $C_{n\varphi n}$. The previous value of n implied the energy spectrum given by eq. 2.7 had decreasing values as n increased due to the sign of the potential energy $1/r^2$ in eq. 2.11. (We absorbed the negativity in the const factor.)

The values of r_n in eq. 2.16 show increasing values for positive increasing n values. However for negative values of n we see r_n decreases with n. This seeming problem is easily handled by changing the value of b which we choose to be $b = \pi/2$ so that r_n decreases from the maximum value corresponding to dimension 18 (which followed from a Consistency Condition found in our earlier books.)

$$b = \pi(18 + ½) \tag{2.21}$$

Then the set of zeros is

$$r'_n = \exp((18 + n')/(\lambda_{n\varphi} - ¼)^{½}) \tag{2.22}$$

where n′ ranges downward from zero to –18. Now the zeros r'_n range downward from

$$r'_0 = \exp(18/(\lambda_{n\varphi} - ¼)^{½})$$

to

$$r'_{18} = 1/(\lambda_{n\varphi} - ¼)^{½})$$

Thus the set of zeros may be viewed corresponding to positive n as above or as corresponding to negative n. *The zeros of the scaling 1/r2 Master Equation can be viewed as occurring for positive radial r coordinate values and non-negative n values as in eq. 2.17.*

Sequence	n_φ	g_g multiplier factor	λ_{n_φ}
?	$7.96 \approx 8$	$\sqrt{2}$	82.42
?	6	$1.585 \approx \frac{1}{2}\pi$	46.778
A1 Cosmos spaces dimension array column lengths $2^{r/2+2}$	4	2	20.79
A2 Cosmos unification space dimension r	4	2	20.79
A3 The sequence of Unification Spaces	4	2	20.79
B Coupling Constants Multiplier	2	4	5.39
C Fermion Sequence 2 Masses	1	$2/3 \times 2^5$	1.3
D Fermion Sequence 1 Masses	$\frac{1}{2}^{\frac{1}{2}}$	$3/2 \times \pi 2^5$	0.65
E n_φ Values Sequence; A Sequence of Sequences	$\frac{1}{2}$	95939	0.325
F New n_φ Values Sequence based on α_{QED}	$0.48 \approx 2^{\frac{1}{2}}/3$	844242	0.303

Figure 2.1. Extended Cosmos Theory Sequences with a sequence determined by setting $\lambda = .303$, which is the value of the QED couping constant e which determines the Fine Structure Constant α. The first row is for the case of a multiplier $\sqrt{2}$. Its n_φ value is an extension of the series of n_φ values 2, 4, 8 of rows A1 through B. The first and second rows are for the cases $n_\varphi = 6$ and $n_\varphi = 8$ which also may be of interest. The second row may correspond to 1.58-bit Large Language Models. See chapter 5.

3. Harmonic Oscillator Form of the Master Equation

We now transform the scaling Master Equation to a harmonic oscillator-like form. Our purpose is to show the Master Equation can accommodate both sequences of powers and linear sequences. The map to harmonic oscillator form also has the advantage of showing the "stability" of the solution to small oscillations. The harmonic oscillator (HO) form of the Master Equation has an equilibrium point for simple harmonic motion with constant frequency and amplitude. Small perturbations often can be approximated by an HO.

3.1 Transformation of Scaling Master Equation to HO Form

The scaling Master Equation has the form

$$(\partial^2/\partial r^2 + \lambda_{n\varphi}/mr^2 - mC)w(r) = 0 \tag{3.1}$$

From eq. 2.11.

If we define a variable s with

$$s = \ln r \tag{3.2}$$

then

$$(\partial^2/\partial s^2 - \partial/\partial s + \lambda_{n\varphi}/m - mCr^2)w(r) = 0 \tag{3.3}$$

Letting

$$w(r) = e^{s/2}\, v(s) = r^{½}\, v(s) \tag{3.4}$$

where, from eq. 2.10, we obtain

$$v(s) = r^{-½}\, w(r) = \psi(r) \tag{3.5}$$

Then we find

$$(\partial^2/\partial s^2 + (\lambda_{n\varphi}/m - ¼) \;\; - mCr^2)v(s) = 0 \tag{3.6}$$

or

$$(\partial^2/\partial s^2 + (\lambda_{n\varphi}/m - ¼) \;\; - mCe^{2s})v(s) = 0 \tag{3.7}$$

The mCe^{2s} term is the potential after the transformation to a harmonic oscillator-like form. We now connect the v(s) solution[6] for small r and thus large negative s to an approximate ψ(r) solution for small r given in eq. 2.14. We approximate eq. 3.7 with

$$(\partial^2/\partial s^2 + \lambda_{n\varphi}/m - ¼\,)v(s) = 0 \tag{3.8}$$

---

[6] We approximate the potential at its minimum relying on the scalability of r. Thus the small r (large negative s) neighborhood represents the potentials well.

Eq. 3.8 is a free equation for v(s). The generality of the Master Equation solution is supported by its solution being equivalent to a Free Field solution (eq. 3.10 below) for small values of the radial parameter r.

A relevant form of the solution for comparison with eq. 2.14 is (using similar constants)

$$v(s) \approx B\cos((\lambda_{n\varphi} - 1/4)^{1/2} s + b) \tag{3.9}$$

Note eq. 2.14 can be expressed as

$$\psi(r) \approx B\cos((\lambda_{n\varphi} - 1/4)^{1/2} s + b) \tag{3.10}$$

Eq. 3.8 corresponds to a free quantum field. There is no direct quantization of the energy. The energy is a positive number. However we can impose a quantization based on a requirement of orthogonality of solutions (an observation due to von Neumann and Case) as we did in the scaling case. Then the energy is quantized due to the form of the solution in eqs. 3.9 and 3.10 for s. The quantized energy $E_{vn\varphi n}$ is

$$E_{vn\varphi n} = n\pi/(\lambda_{n\varphi} - 1/4)^{1/2} \tag{3.11}$$

It agrees with the scaling Master Equation quantization of eq. 2.8:

$$C_{n\varphi n} = c_{n\varphi} \exp(n\pi/(\lambda_{n\varphi} - 1/4)^{1/2}) = c_{n\varphi} \exp(E_{vn\varphi n}) \tag{3.12}$$

The constant n can be positive or negative. We choose n to be positive with n and n_φ being independent quantum numbers.

We thus find agreement between this approximate harmonic oscillator form of the Master Equation and the scaling Master Equation.

The transformation to an approximate harmonic oscillator form implies a form of stability for the Master Equation solutions. It also suggests a universality due to the generality of harmonic oscillator solutions under small perturbations for many Physical forces.

3.2 Equally Spaced Harmonic Oscillator-like Eigenvalues

The solution for ψ as a function of s displays equally spaced harmonic oscillator eigenvalues. Using eq. 3.10 we find the zeros of ψ satisfy the analogue of eq. 2.15:

$$\begin{aligned}(\lambda_{n\varphi} - 1/4)^{1/2} s_n + b &= \pm\pi/2, \pm 3\pi/2, \pm 5\pi\ 2, \pm 7\pi/2m\ ,,, \\ &= (n + 1/2)\,\pi\end{aligned} \tag{3.13}$$

or

$$s_n = ((n + 1/2)\,\pi - b)/(\lambda_{n\varphi} - 1/4)^{1/2} \tag{3.14}$$

Setting b = ½π results in an equally spaced Harmonic oscillator-like spectrum of zeros:

$$s_n = n\pi/(\lambda_{n\varphi} - \tfrac{1}{4})^{1/2} \qquad (3.15)$$

with

$$r_n = \exp(s_n) \qquad (3.15a)$$

which corresponds to eq. 2.17.

In deriving eq. 2.13 we obtained a seemingly positive value for s_n. However the approximation eq. 3.9 must apply for negative s values. Thus a superficial dilemma appears. It can be resolved in a manner similar to the scaling Master Equation case in section 2.8.

We redefine the constant b to be equal to the maximum value of n = 18 (the Cosmos dimension found in the author's Consistency Condition):

$$b = \pi(18 + \tfrac{1}{2}) \qquad (3.16)$$

then from eq. 3.14 we find

$$s_n = (n - 18)\pi/(\lambda_{n\varphi} - \tfrac{1}{4})^{1/2} < 0 \qquad (3.17)$$

giving

$$s_0 = -18\pi/(\lambda_{n\varphi} - \tfrac{1}{4})^{1/2} < 0 \qquad (3.18)$$

and

$$s_{18} = 0 \qquad (3.19)$$

The approximate harmonic oscillator Master Equation implies the same spectrum and corresponding function zeros as the scaling Master Equation. The Master Equation has the desired scaling and fractal features. It is complemented by the harmonic oscillator-like equivalent.

We have found an approximate Harmonic Oscillator-like form of the Master Equation. It supports the generality of the Master Equation.

4. The Point - Master Equation Connection

The primeval Mathematical point leading to the Fractal unit square and to a two dimension space superficially appears to be disjoint from the Master Equation. The Master Equation generates a progression of structure that leads fairly directly to coupling constant values and fundamental fermion masses.

At first glance the progression from Mathematical point to two dimension space does not appear to have experimental support. Thus there is a perceived conceptual "gap" between the progression and the Master Equation.

There are a number of reasons supporting the connection of the progression to the Master Equation:

1. The Master Equation solutions have zeros that can be put into a one to one correspondence with the piecewise linear line segments used in the construction of the fractal Cosmos curve that leads to the 2D space for the Master Equation. The values of $r_n = 2^n$ in eq. 2.20 specify the length of the piecewise linear line segments that generate a Cosmos (and Hilbert) fractal curve. Other choices of the $g_{n\varphi}$ value in eq. 2.19 generate other power sequences that can be scaled to the Hilbert sequence powers of 2^n. One can view the Master Equation as implying the fractal construction process. Thus the construction process from a point to the fractal-based unit square can be viewed as embodied in the Master Equation. The fractal construction process has an integral part in the generation of Cosmos structure.

2. The Master Equation solutions can be viewed as defining a generating function for the lengths of the piecewise linear line segments of the fractal.

3. The Master Equation has features that make it the simplest, and possibly the most general, choice of possible Cosmos generating equations:

    a. First it is a second order differential equation similar to the two dimension Schrödinger equation.
    b. It has a $1/r^2$ potential that enables the avoidance of the introduction of a mass scale.
    c. It can be transformed to a harmonic oscillator-like form with an equally spaced eigenvalue spectrum.
    d. The generality of the scaling Master Equation solution is supported by its solution also being equivalent to a Free Field solution for small values of the radial parameter.

4. The fractal curve construction supports a map between the 1D piecewise linear segments and the 2D unit square and space that preserves locality: two data points that are close to each other in a 1D line segment are close to each other after the line is folded in the fractal construction process. Locality is a very desirable feature in Physics.

5. The Master Equation, which embodies scaling, can be transformed to a harmonic oscillator-like form with a corresponding eigenvalue spectrum. This form, which characterizes small perturbative forces, provides a type of stability against disturbances from small forces.

6. The generality of the scaling Master Equation solution is supported by its solution also being equivalent to a Free Field solution for small values of the radial parameter r.

7. We have shown from the Master Equation solution that the evolutionary sequence of the Cosmos follows. We see it encompasses the initial mathematical point, the generation of a two dimension piecewise linear unit length, the generation of the fractal Cosmos Curve, the generation of the Cosmos spaces and their dimension arrays, the unification spaces sequence, and the set of sequences of Fig. 2.1.

The Master Equation is the basis of the evolution of the Cosmos conceptually. As we have often pointed out, Cosmos evolution is instantaneous in the sense that there is no time coordinate. Therefore it is timeless! The Master Equation embodies the Cosmos and sets the stage for its universes. The unity of the progression from point to Master Equation is established.

5. 1.58-bit Large Model Language Approach

Fig. 2.1 revealed the interesting value of g_g =1.585 ≈ π/2 for $n_φ$ = 6. This value is approximately the 1.58 value in the 1.58-bit Ternary Large Language Model (LLM). The ternary LLM is based on the trit – a triplet of data values (-1, 0, 1). The 1.58 trit emerges from

$$\log_2 3 = 1.58 \text{ bits} \tag{5.1}$$

The use of trits for data storage gives a significant compression of data. If we compare a trit formulation with the Cosmos dimension array size of 2^{r+4} for a space of dimension rwe find

Setting Dimension array size $2^{r+4} = 3^s$ for data storage

where s is a trit size. Then we obtain

$$r+4 = s \log_2 3 = 1.58s$$

An increase of r by 2 to next dimension array (data warehouse) gives a change in LLM size

$$2 = 1.58t$$
$$t = 2/1.58 = 1.27$$

giving a data storage that is now $3^{s+1.27}$ compare to increase in r to 2^{r+6}. The result, in this example, is a data compression by a factor of (2 – 1.27)/2 = 37% in data storage.

1.58-bit LLMs have the features:

Saves on computation
Reduced computation: Replaces multiplies with adds
Reduces storage memory

The 2^{r+4} dimension arrays does have the advantage of a clean connection to the fundamental fermion spectrum and the set of internal symmetries.

5.1 LLM and Fractal Curve Locality

LLM data storage preserves locality of the data. A fractal Hilbert-like curve, such as that of Cosmos Theory, has a similar locality property. Both have a number of computation applications:

IP addresses mapping,

robots,
location spacing data,
data warehouses,
dithering,
AI applications

We thus see Hilbert curves such as our Cosmos Curve have LLM-like features. A most economical theory!

5.2 Is there a deeper level based on compactness of data stores?

Cosmos Theory – based as it is on a Fractal curve – exhibits the locality found in LLMs. Then does our Cosmos Theories form an LLM simulation of Reality? A question worth pursuing!

6. General Considerations on the Construction of the Cosmos

One Cosmos suffices.

Two questions confront us as a result of our Cosmos Theory development:

Why a Cosmos?

Why the form of the Cosmos?

The first question does not seem to have a Physical answer. The second question may have a Physical answer based on the information that we have gleaned from our investigation of the features and aspects of Cosmos Theory (including Sequence Theory and Gambol Theory).

6.1 Simplicity

A true theory describing the origin of the Cosmos should not have a pre-history. Otherwise one must find the origin of the pre-history – possibly resulting in an unending, widening circle of theory.

6.2 Generality

A true theory must encompass all Physical aspects of the Cosmos. Piecemeal partial solutions are not acceptable.

6.3 Scaling and Other Parameters

Scaling and other parameters of an *ad hoc* nature should be avoided. Otherwise they also would require justification.

6.4 Supports a Pattern of Growth

A true theory of the Cosmos should have a pattern of growth that enables it to encompass all aspects of Physics. Euclidean Geometry is an example of a growable theory with a pattern of growth that takes fundamental features to the complex such as complex geometric structures.

6.5 Good Structuring

We often hear of Nature reusing successful features. A true theory of the Cosmos must support reusability. Our Cosmos Theory obtained good reusability by separating structure (the form of the set of spaces) from substance with the implementation of substance in the form of universes.

6.6 Direct Formulation of Features from Nothing

A true theory must start from "nothing" excluding everything (including a vacuum which after all presumes structure). Our Cosmos Theory starts from a dimensionless point and precedes though a construction process to a two dimension space.

6.7 Key Central Formulation

A true theory has a descriptive central formulation that generates the major features of the theory. The formulation must dovetail with the preceding part that leads to its formulation. In Cosmos Theory there is a Master Equation, which we have seen is related to the fractal growth of the two dimension space. The Master Equation in both its scaling and harmonic oscillator-like forms directly gives the core set of sequences that govern the set of spaces, the coupling constants, the fermion masses, and other UST features.

The Master Equation is a differential equation with the features:

`1. A second order Schrödinger-like quantum mechanical differential.
(Space-time and Relativity do not exist in the two dimension space.)
2. A scaling $1/r^2$ potential that prevents the appearance of a mass scale.
3. An energy term that is quantized.
4. The avoidance of other terms of no apparent consequence.
A most minimal differential equation.
5. A corresponding harmonic oscillator-like formulation that implies a form of stability seen in classical harmonic oscillator studies. In classical Physics many forces are approximated by harmonic oscillator behavior for small perturbations.
6. The generality of the scaling Master Equation solution is supported by its solution also being equivalent to a Free Field solution for small values of the radial parameter r.

6.8 Economy in Theory, Storage and Computation

The Cosmos Theory originates in a fractal curve that supports locality in the same manner as Large Language Models. See Chapter 5. The result is an economy in data (particle) storage and a facility in computation. Perturbation Theory is more tractable than complex fluid dynamics calculations.

The form of Cosmos Theory is evidence of its theoretic economy as described above.

We have thus come to an understanding of key points in the construction of a theory of the Cosmos and their reality in our Cosmos Theory.

REFERENCES

Akhiezer, N. I., Frink, A. H. (tr), 1962, *The Calculus of Variations* (Blaisdell Publishing, New York, 1962).
Bjorken, J. D., Drell, S. D., 1964, *Relativistic Quantum Mechanics* (McGraw-Hill, New York, 1965).
Bjorken, J. D., Drell, S. D., 1965, *Relativistic Quantum Fields* (McGraw-Hill, New York, 1965).
Blaha, S., 1995, *C++ for Professional Programming* (International Thomson Publishing, Boston, 1995).
_______, 1998, *Cosmos and Consciousness* (Pingree-Hill Publishing, Auburn, NH, 1998 and 2002).
_______, 2002, *A Finite Unified Quantum Field Theory of the Elementary Particle Standard Model and Quantum Gravity Based on New Quantum Dimensions™ & a New Paradigm in the Calculus of Variations* (Pingree-Hill Publishing, Auburn, NH, 2002).
_______, 2004, *Quantum Big Bang Cosmology: Complex Space-time General Relativity, Quantum Coordinates™ Dodecahedral Universe, Inflation, and New Spin 0, ½, 1 & 2 Tachyons & Imagyons* (Pingree-Hill Publishing, Auburn, NH, 2004).
______, *2005a, Quantum Theory of the Third Kind: A New Type of Divergence-free Quantum Field Theory Supporting a Unified Standard Model of Elementary Particles and Quantum Gravity based on a New Method in the Calculus of Variations* (Pingree-Hill Publishing, Auburn, NH, 2005).
______, 2005b, *The Metatheory of Physics Theories, and the Theory of Everything as a Quantum Computer Language* (Pingree-Hill Publishing, Auburn, NH, 2005).
______, 2005c, *The Equivalence of Elementary Particle Theories and Computer Languages: Quantum Computers, Turing Machines, Standard Model, Superstring Theory, and a Proof that Gödel's Theorem Implies Nature Must Be Quantum* (Pingree-Hill Publishing, Auburn, NH, 2005).
______, 2006a, *The Foundation of the Forces of Nature* (Pingree-Hill Publishing, Auburn, NH, 2006).
______, 2006b, *A Derivation of ElectroWeak Theory based on an Extension of Special Relativity; Black Hole Tachyons; & Tachyons of Any Spin.* (Pingree-Hill Publishing, Auburn, NH, 2006).
_______, 2007a, *Physics Beyond the Light Barrier: The Source of Parity Violation, Tachyons, and A Derivation of Standard Model Features* (Pingree-Hill Publishing, Auburn, NH, 2007).
______, 2007b, *The Origin of the Standard Model: The Genesis of Four Quark and Lepton Species, Parity Violation, the ElectroWeak Sector, Color SU(3), Three Visible Generations of Fermions, and One Generation of Dark Matter with Dark Energy* (Pingree-Hill Publishing, Auburn, NH, 2007).
______, *2008a, A Direct Derivation of the Form of the Standard Model From GL(16) (Pingree-Hill Publishing, Auburn, NH, 2008).*
______, 2008b, *A Complete Derivation of the Form of the Standard Model With a New Method to Generate Particle Masses Second Edition* (Pingree-Hill Publishing, Auburn, NH, 2008)
______, 2009, *The Algebra of Thought & Reality: The Mathematical Basis for Plato's Theory of Ideas, and Reality Extended to Include A Priori Observers and Space-Time Second Edition* (Pingree-Hill Publishing, Auburn, NH, 2009).
______, 2009a, *Bright Stars, Bright Universe* (Pingree-Hill Publishing, Auburn, NH, 2008)
______, 2010a, *Operator Metaphysics: A New Metaphysics Based on a New Operator Logic and a New Quantum Operator Logic that Lead to a Mathematical Basis for Plato's Theory of Ideas and Reality* (Pingree-Hill Publishing, Auburn, NH, 2010).
______, 2010b, *The Standard Model's Form Derived from Operator Logic, Superluminal Transformations and GL(16)* (Pingree-Hill Publishing, Auburn, NH, 2010).
______, 2010c, *SuperCivilizations: Civilizations as Superorganisms* (McMann-Fisher Publishing, Auburn, NH, 2010).
______, 2011a, *21st Century Natural Philosophy Of Ultimate Physical Reality* (McMann-Fisher Publishing, Auburn, NH, 2011).
______, 2011b, *All the Universe! Faster Than Light Tachyon Quark Starships & Particle Accelerators with the LHC as a Prototype Starship Drive Scientific Edition* (Pingree-Hill Publishing, Auburn, NH, 2011).
______, 2011c, *From Asynchronous Logic to The Standard Model to Superflight to the Stars* (Blaha Research, Auburn, NH, 2011).

______, 2012a, *From Asynchronous Logic to The Standard Model to Superflight to the Stars volume 2: Superluminal CP and CPT, U(4) Complex General Relativity and The Standard Model, Complex Vierbein General Relativity, Kinetic Theory, Thermodynamics* (Blaha Research, Auburn, NH, 2012).
______, 2012b, *Standard Model Symmetries, And Four And Sixteen Dimension Complex Relativity; The Origin Of Higgs Mass Terms* (Blaha Reasearch, Auburn, NH, 2012).
______, 2013a, *Multi-Stage Space Guns, Micro-Pulse Nuclear Rockets, and Faster-Than-Light Quark-Gluon Ion Drive Starships* (Blaha Research, Auburn, NH, 2013).
______, 2013b, *The Bridge to Dark Matter; A New Sibling Universe; Dark Energy; Inflatons; Quantum Big Bang; Superluminal Physics; An Extended Standard Model Based on Geometry* (Blaha Reasearch, Auburn, NH, 2013).
______, 2014a, *Universes and Megaverses: From a New Standard Model to a Physical Megaverse; The Big Bang; Our Sibling Universe's Wormhole; Origin of the Cosmological Constant, Spatial Asymmetry of the Universe, and its Web of Galaxies; A Baryonic Field between Universes and Particles; Megaverse Extended Wheeler-DeWitt Equation* (Blaha Reasearch, Auburn, NH, 2014).
______, 2014b, *All the Megaverse! Starships Exploring the Endless Universes of the Cosmos Using the Baryonic Force* (Blaha Research, Auburn, NH, 2014).
______, 2014c, *All the Megaverse! II Between Megaverse Universes: Quantum Entanglement Explained by the Megaverse Coherent Baryonic Radiation Devices – PHASERs Neutron Star Megaverse Slingshot Dynamics Spiritual and UFO Events, and the Megaverse Microscopic Entry into the Megaverse* (Blaha Research, Auburn, NH, 2014).
______, 2015a, *PHYSICS IS LOGIC PAINTED ON THE VOID: Origin of Bare Masses and The Standard Model in Logic, U(4) Origin of the Generations, Normal and Dark Baryonic Forces, Dark Matter, Dark Energy, The Big Bang, Complex General Relativity, A Megaverse of Universe Particles* (Blaha Research, Auburn, NH, 2015).
______, 2015b, *PHYSICS IS LOGIC Part II: The Theory of Everything, The Megaverse Theory of Everything, U(4)⊗U(4) Grand Unified Theory (GUT), Inertial Mass = Gravitational Mass, Unified Extended Standard Model and a New Complex General Relativity with Higgs Particles, Generation Group Higgs Particles* (Blaha Research, Auburn, NH, 2015).
____, 2015c, *The Origin of Higgs ("God") Particles and the Higgs Mechanism: Physics is Logic III, Beyond Higgs – A Revamped Theory With a Local Arrow of Time, The Theory of Everything Enhanced, Why Inertial Frames are Special, Universes of the Mind* (Blaha Research, Auburn, NH, 2015).
______, 2015d, *The Origin of the Eight Coupling Constants of The Theory of Everything: U(8) Grand Unified Theory of Everything (GUTE), S^8 Coupling Constant Symmetry, Space-Time Dependent Coupling Constants, Big Bang Vacuum Coupling Constants, Physics is Logic IV* (Blaha Research, Auburn, NH, 2015).
______, 2016a, *New Types of Dark Matter, Big Bang Equipartition, and A New U(4) Symmetry in the Theory of Everything: Equipartition Principle for Fermions, Matter is 83.33% Dark, Penetrating the Veil of the Big Bang, Explicit QFT Quark Confinement and Charmonium, Physics is Logic V* (Blaha Research, Auburn, NH, 2016).
______, 2016b, *The Periodic Table of the 192 Quarks and Leptons in The Theory of Everything: The U(4) Layer Group, Physics is Logic VI* (Blaha Research, Auburn, NH, 2016).
______, 2016c, *New Boson Quantum Field Theory, Dark Matter Dynamics, Dark Matter Fermion Layer Mixing, Genesis of Higgs Particles, New Layer Higgs Masses, Higgs Coupling Constants, Non-Abelian Higgs Gauge Fields, Physics is Logic VII* (Blaha Research, Auburn, NH, 2016).
______, 2016d, *Unification of the Strong Interactions and Gravitation: Quark Confinement Linked to Modified Short-Distance Gravity; Physics is Logic VIII* (Blaha Research, Auburn, NH, 2016).
______, 2016e, *MoND: Unification of the Strong Interactions and Gravitation II, Quark Confinement Linked to Large-Scale Gravity, Physics is Logic IX* (Blaha Research, Auburn, NH, 2016).
______, 2016f, *CQ Mechanics: A Unification of Quantum & Classical Mechanics, Quantum/Semi-Classical Entanglement, Quantum/Classical Path Integrals, Quantum/Classical Chaos* (Blaha Research, Auburn, NH, 2016).
______, 2016g, *GEMS Unified Gravity, ElectroMagnetic and Strong Interactions: Manifest Quark Confinement, A Solution for the Proton Spin Puzzle, Modified Gravity on the Galactic Scale* (Pingree Hill Publishing, Auburn, NH, 2016).
______, 2016h, *Unification of the Seven Boson Interactions based on the Riemann-Christoffel Curvature Tensor* (Pingree Hill Publishing, Auburn, NH, 2016).
______, 2017a, *Unification of the Eleven Boson Interactions based on 'Rotations of Interactions'* (Pingree Hill Publishing, Auburn, NH, 2017).
______, 2017b, *The Origin of Fermions and Bosons, and Their Unification* (Pingree Hill Publishing, Auburn, NH, 2017).
______, 2017c, *Megaverse: The Universe of Universes* (Pingree Hill Publishing, Auburn, NH, 2017).
______, 2017d, *SuperSymmetry and the Unified SuperStandard Model* (Pingree Hill Publishing, Auburn, NH, 2017).

______, 2017e, *From Qubits to the Unified SuperStandard Model with Embedded SuperStrings: A Derivation* (Pingree Hill Publishing, Auburn, NH, 2017).
______, 2017f, *The Unified SuperStandard Model in Our Universe and the Megaverse: Quarks, ...* , (Pingree Hill Publishing, Auburn, NH, 2017).
______, 2018a, *The Unified SuperStandard Model and the Megaverse SECOND EDITION A Deeper Theory based on a New Particle Functional Space that Explicates Quantum Entanglement Spookiness (Volume 1)* (Pingree Hill Publishing, Auburn, NH, 2018).
______, 2018b, *Cosmos Creation: The Unified SuperStandard Model, Volume 2, SECOND EDITION* (Pingree Hill Publishing, Auburn, NH, 2018).
______, 2018c, *God Theory (*Pingree Hill Publishing, Auburn, NH, 2018).
______, 2018d, *Immortal Eye: God Theory: Second Edition* (Pingree Hill Publishing, Auburn, NH, 2018).
______, 2018e, *Unification of God Theory and Unified SuperStandard Model THIRD EDITION* (Pingree Hill Publishing, Auburn, NH, 2018).
______, 2019a, *Calculation of: QED α = 1/137, and Other Coupling Constants of the Unified SuperStandard Theory* (Pingree Hill Publishing, Auburn, NH, 2019).
______, 2019b, *Coupling Constants of the Unified SuperStandard Theory SECOND EDITION* (Pingree Hill Publishing, Auburn, NH, 2019).
______, 2019c, *New Hybrid Quantum Big_Bang–Megaverse_Driven Universe with a Finite Big Bang and an Increasing Hubble Constant* (Pingree Hill Publishing, Auburn, NH, 2019).
______, 2019d, *The Universe, The Electron and The Vacuum* (Pingree Hill Publishing, Auburn, NH, 2019).
______, 2019e, *Quantum Big Bang – Quantum Vacuum Universes (Particles)* (Pingree Hill Publishing, Auburn, NH, 2019).
______, 2019f, *The Exact QED Calculation of the Fine Structure Constant Implies ALL 4D Universes have the Same Physics/Life Prospects* (Pingree Hill Publishing, Auburn, NH, 2019).
______, 2019g, *Unified SuperStandard Theory and the SuperUniverse Model: The Foundation of Science* (Pingree Hill Publishing, Auburn, NH, 2019).
______, 2020a, *Quaternion Unified SuperStandard Theory (The QUeST) and Megaverse Octonion SuperStandard Theory (MOST)* (Pingree Hill Publishing, Auburn, NH, 2020).
______, 2020b, *United Universes Quaternion Universe - Octonion Megaverse* (Pingree Hill Publishing, Auburn, NH, 2020).
______, 2020c, *Unified SuperStandard Theories for Quaternion Universes & The Octonion Megaverse* (Pingree Hill Publishing, Auburn, NH, 2020).
______, 2020d, *The Essence of Eternity: Quaternion & Octonion SuperStandard Theories* (Pingree Hill Publishing, Auburn, NH, 2020).
______, 2020e, *The Essence of Eternity II* (Pingree Hill Publishing, Auburn, NH, 2020).
______, 2020f, *A Very Conscious Universe* (Pingree Hill Publishing, Auburn, NH, 2020).
______, 2020g, *Hypercomplex Universe* (Pingree Hill Publishing, Auburn, NH, 2020).
______, 2020h, *Beneath the Quaternion Universe* (Pingree Hill Publishing, Auburn, NH, 2020).
______, 2020i, *Why is the Universe Real? From Quaternion & Octonion to Real Coordinates* (Pingree Hill Publishing, Auburn, NH, 2020).
______, 2020j, *The Origin of Universes: of Quaternion Unified SuperStandard Theory (QUeST); and of the Octonion Megaverse (UTMOST)* (Pingree Hill Publishing, Auburn, NH, 2020).
______, 2020k, *The Seven Spaces of Creation: Octonion Cosmology* (Pingree Hill Publishing, Auburn, NH, 2020).
______, 2020l, *From Octonion Cosmology to the Unified SuperStandard Theory of Particles* (Pingree Hill Publishing, Auburn, NH, 2020).
______, 2021a, *Pioneering the Cosmos* (Pingree Hill Publishing, Auburn, NH, 2021).
______, 2021b, *Pioneering the Cosmos II* (Pingree Hill Publishing, Auburn, NH, 2021).
______, 2021c, *Beyond Octonion Cosmology* (Pingree Hill Publishing, Auburn, NH, 2021).
______, 2021d, *Universes are Particles* (Pingree Hill Publishing, Auburn, NH, 2021).
______, 2021e, *Octonion-like dna-based life, Universe expansion is decay, Emerging New Physics* (Pingree Hill Publishing, Auburn, NH, 2021).
______, 2021f, *The Science of Creation New Quantum Field Theory of Spaces* (Pingree Hill Publishing, Auburn, NH, 2021).
______, 2021g, *Quantum Space Theory With Application to Octonion Cosmology & Possibly To Fermionic Condensed Matter* (Pingree Hill Publishing, Auburn, NH, 2021).
______, 2021h, *21st Century Natural Philosophy of Octonion Cosmology , and Predestination, Fate, and Free Will* (Pingree Hill Publishing, Auburn, NH, 2021).

______, 2021i, *Beyond Octonion Cosmology II : Origin of the Quantum; A New Generalized Field Theory (GiFT); A Proof of the Spectrum of Universes; Atoms in Higher Universes* (Pingree Hill Publishing, Auburn, NH, 2021).
______, 2021j, *Integration of General Relativity and Quantum Theory: Octonion Cosmology, GiFT, Creation/Annihilation Spaces CASe, Reduction of Spaces to a Few Fermions and Symmetries in Fundamental Frames* (Pingree Hill Publishing, Auburn, NH, 2021).
______, 2022a, *New View of Octonion Cosmology Based on the Unification of General Relativity and Quantum Theory* (Pingree Hill Publishing, Auburn, NH, 2022).
______, 2022b, *The Dust Beneath Hypercomplex Cosmology* (Pingree Hill Publishing, Auburn, NH, 2022).
______, 2022c, *Passing Through Nature to Eternity: ProtoCosmos, HyperCosmos, Unified SuperStandard Theory* (Pingree Hill Publishing, Auburn, NH, 2022).
______, 2022d, *HyperCosmos Fractionation and Fundamental Reference Frame Based Unification: Particle Inner Space Basis of Parton and Dual Resonance Models* (Pingree Hill Publishing, Auburn, NH, 2022).
______, 2022e, *A New UniDimension ProtoCosmos and SuperString F-Theory Relation to the HyperCosmos* (Pingree Hill Publishing, Auburn, NH, 2022).
______, 2022f, *The Cosmic Panorama: ProtoCosmos, HyperCosmos,Unified SuperStandard Theory (UST) Derivation* (Pingree Hill Publishing, Auburn, NH, 2022).
______, 2022g, *Ultimate Origin: ProtoCosmos and HyperCosmos* (Pingree Hill Publishing, Auburn, NH, 2022).
______, 2023a, *UltraUnification and the Generation of the Cosmos* (Pingree Hill Publishing, Auburn, NH, 2023).
______, 2023b, *God and and Cosmos Theory* (Pingree Hill Publishing, Auburn, NH, 2023).
______, 2023c, *A New Completely Geometric SU(8) Cosmos Theory; New PseudoFermion Fields; Fibonacci-like Dimension Arrays; Ramsey Number Approximation* (Pingree Hill Publishing, Auburn, NH, 2023).
______, 2023d, *Newton's Apple is Now the Fermion* (Pingree Hill Publishing, Auburn, NH, 2023).
______, 2023e,*Cosmos Theory: The Sub-Particle Gambol Model* (Pingree Hill Publishing, Auburn, NH, 2023).
______, 2024a, *Cosmos-Universe-Particle-Gambol Theory* (Pingree Hill Publishing, Auburn, NH, 2024).
______, 2024b, *Fractal Cosmos Theory* (Pingree Hill Publishing, Auburn, NH, 2024).
______, 2024c, *Fractal Cosmic Curve: Tensor-Based CosmosTheory* (Pingree Hill Publishing, Auburn, NH, 2024).
______, 2024d, *The Eternal Form of Cosmos Theory* (Pingree Hill Publishing, Auburn, NH, 2024).
______, 2024e, *The Eternal Form of Cosmos Theory Third Edition* (Pingree Hill Publishing, Auburn, NH, 2024).
______, 2024f, *Fundamental Constants of Cosmos Theory and The Standard Model* (Pingree Hill Publishing, Auburn, NH, 2024).
______, 2024g, *Quark, Lepton, W and Z Masses of Cosmos Theory and The Standard Model* (Pingree Hill Publishing, Auburn, NH, 2024).
______, 2024h, *Geometric Cosmos Geometric Universe* (Pingree Hill Publishing, Auburn, NH, 2024).
______, 2024i, *Particles and Universes of Cosmos Theory* (Pingree Hill Publishing, Auburn, NH, 2024).
______, 2024j, *Unification of the Subluminal and the Superluminal in Cosmos Theory* (Pingree Hill Publishing, Auburn, NH, 2024).
______, 2024k, *The Dawn of Dynamic Cosmos Dimension Arrays* (Pingree Hill Publishing, Auburn, NH, 2024).
______, 2024ℓ, *Structure and Dynamics of Cosmos Theory and the Unified SuperStandard Theory* (Pingree Hill Publishing, Auburn, NH, 2024).
______, 2025a, *Black Holes, White Holes, and Superluminal Starship* (Pingree Hill Publishing, Auburn, NH, 2025).
______, 2025b, *The Sequence Theory Foundation of Cosmos Theory and Gambol Theory* (Pingree Hill Publishing, Auburn, NH, 2025).
____, 2025c, *BEYOND TIME AND SPACE Fermion Strata & Consistency Conditions for Sequences, Cosmos Spaces and Gambols* (Pingree Hill Publishing, Auburn, NH, 2025).
____, 2025d, *The C o s m o s as a Physics Organism* (Blaha Research, Auburn, NH, 2025).
Eddington, A. S., 1952, *The Mathematical Theory of Relativity* (Cambridge University Press, Cambridge, U.K., 1952).
Fant, Karl M., 2005, *Logically Determined Design: Clockless System Design With NULL Convention Logic* (John Wiley and Sons, Hoboken, NJ, 2005).
Feinberg, G. and Shapiro, R., 1980, *Life Beyond Earth: The Intelligent Earthlings Guide to Life in the Universe* (William Morrow and Company, New York, 1980).
Gelfand, I. M., Fomin, S. V., Silverman, R. A. (tr), 2000, *Calculus of Variations* (Dover Publications, Mineola, NY, 2000).
Giaquinta, M., Modica, G., Souchek, J., 1998, *Cartesian Coordinates in the Calculus of Variations* Volumes I and II (Springer-Verlag, New York, 1998).
Giaquinta, M., Hildebrandt, S., 1996, *Calculus of Variations* Volumes I and II (Springer-Verlag, New York, 1996).

Gradshteyn, I. S. and Ryzhik, I. M., 1965, *Table of Integrals, Series, and Products* (Academic Press, New York, 1965).
Heitler, W., 1954, *The Quantum Theory of Radiation* (Claendon Press, Oxford, UK, 1954).
Huang, Kerson, 1992, *Quarks, Leptons & Gauge Fields 2nd Edition* (World Scientific Publishing Company, Singapore, 1992).
Jost, J., Li-Jost, X., 1998, *Calculus of Variations* (Cambridge University Press, New York, 1998).
Kaku, Michio, 1993, *Quantum Field Theory*, (Oxford University Press, New York, 1993).
Kirk, G. S. and Raven, J. E., 1962, *The Presocratic Philosophers* (Cambridge University Press, New York, 1962).
Landau, L. D. and Lifshitz, E. M., 1987, *Fluid Mechanics 2nd Edition*, (Pergamon Press, Elmsford, NY, 1987).
Rescher, N., 1967, *The Philosophy of Leibniz* (Prentice-Hall, Englewood Cliffs, NJ, 1967).
Riesz, Frigyes and Sz.-Nagy, Béla, 1990, *Functional Analysis* (Dover Publications, New York, 1990).
Sakurai, J. J., 1964, *Invariance Principles and Elementary Particles* (Princeton University Press, Princeton, NJ, 1964).
Weinberg, S., 1972, *Gravitation and Cosmology* (John Wiley and Sons, New York, 1972).
Weinberg, S., 1995, *The Quantum Theory of Fields Volume I* (Cambridge University Press, New York, 1995).

INDEX

About the Author

Stephen Blaha is a well-known Physicist and Man of Letters with interests in Science, Society and civilization, the Arts, and Technology. He had an Alfred P. Sloan Foundation scholarship in college. He received his Ph.D. in Physics from Rockefeller University. He has served on the faculties of several major universities. He was also a Member of the Technical Staff at Bell Laboratories, a manager at the Boston Globe Newspaper, a Director at Wang Laboratories, and President of Blaha Software Inc. and of Janus Associates Inc. (NH).

Among other achievements he was a co-discoverer of the "r potential" for heavy quark binding developing the first (and still the only demonstrable) non-Aeolian gauge theory with an "r" potential; first suggested the existence of topological structures in superfluid He-3; first proposed Yang-Mills theories would appear in condensed matter phenomena with non-scalar order parameters; first developed a grammar-based formalism for quantum computers and applied it to elementary particle theories; first developed a new form of quantum field theory without divergences (thus solving a major 60 year old problem that enabled a unified theory of the Standard Model and Quantum Gravity without divergences to be developed); first developed a formulation of complex General Relativity based on analytic continuation from real space-time; first developed a generalized non-homogeneous Robertson-Walker metric that enabled a quantum theory of the Big Bang to be developed without singularities at t = 0; first generalized Cauchy's theorem and Gauss' theorem to complex, curved multi-dimensional spaces; received Honorable Mention in the Gravity Research Foundation Essay Competition in 1978; first developed a physically acceptable theory of faster-than-light particles; first derived a composition of extremums method in the Calculus of Variations; first quantitatively suggested that inflationary periods in the history of the universe were not needed; first proved Gödel's Theorem implies Nature must be quantum; provided a new alternative to the Higgs Mechanism, and Higgs particles, to generate masses; first showed how to resolve logical paradoxes including Gödel's Undecidability Theorem by developing Operator Logic and Quantum Operator Logic; first developed a quantitative harmonic oscillator-like model of the life cycle, and interactions, of civilizations; first showed how equations describing superorganisms also apply to civilizations. A recent book shows his theory applies successfully to the past 14 years of history and to *new* archaeological data on Andean and Mayan civilizations as well as Early Anatolian and Egyptian civilizations.

He first developed an axiomatic derivation of the form of The Standard Model from geometry – space-time properties – The Unified SuperStandard Model. It unifies all the known forces of Nature. It also has a Dark Matter sector that includes a Dark ElectroWeak sector with Dark doublets and Dark gauge interactions. It uses quantum coordinates to remove infinities that crop up in most interacting quantum field theories and additionally to remove the infinities that appear in the Big Bang and generate inflationary growth of the universe. It shows gravity has a MOND-like form without sacrificing Newton's Laws. It relates the interactions of the MOND-like sector of gravity with the r-potential of Quark Confinement. The axioms of the theory lead to the question of their origin. We suggest in the preceding edition of this book it can be attributed to an entity with God-like properties. We explore these properties in "God Theory" and show they predict that the Cosmos exists forever although individual universes (or incarnations of our universe) "come and go." Several other important results emerge from God Theory such a functionally triune God. The Unified SuperStandard Theory has many other important parts described in the Current Edition of *The Unified SuperStandard Theory* and expanded in subsequent volumes.

Blaha has had a major impact on a succession of elementary particle theories: his Ph.D. thesis (1970), and papers, showed that quantum field theory calculations to all orders in ladder approximations could not give scaling deep inelastic electron-nucleon scattering. He later showed the eigenvalue equation for the fine structure constant α in Johnson-Baker-Willey QED had a zero at $\alpha = 1$ not 1/137 by solving the Schwinger-Dyson equations to all orders in an approximation that agreed with exact results to 4th order in α thus ending interest in this theory. In 1979 at Prof. Ken Johnson's (MIT) suggestion he calculated the proton-neutron mass difference in the MIT bag model and found the result had the wrong sign reducing interest in the bag model. These results all appear in Physical Review papers. In the 2000's he repeatedly pointed out the shortcomings of SuperString theory and showed that The Standard Model's form could be derived from space-time geometry by an extension of Lorentz transformations to faster than light transformations. This deeper space-time basis greatly increases the possibility that it is part of THE fundamental theory. Recently, Blaha showed that the Weak interactions differed significantly from the Strong, electromagnetic and gravitation interactions in important respects while these interactions had similar features, and suggested that ElectroWeak theory, which is essentially a glued union of the Weak interactions and Electromagnetism, possibly modulo unknown Higgs particle features, be replaced by a unified theory of the other interactions combined with a stand-alone Weak interaction theory. Blaha also showed that, if Charmonium calculations are taken seriously, the Strong interaction coupling constant is only a factor of five larger than the electromagnetic coupling constant, and thus Strong interaction perturbation theory would make sense and yield physically meaningful results.

In graduate school (1965-71) he wrote substantial papers in elementary particles and group theory: The Inelastic E- P Structure Functions in a Gluon Model. Phys. Lett. B40:501-502,1972; Deep-Inelastic E-P Structure Functions In A Ladder Model With Spin 1/2 Nucleons, Phys.Rev. D3:510-523,1971; Continuum Contributions To The Pion Radius, Phys. Rev. 178:2167-2169,1969; Character Analysis of U(N) and SU(N), J. Math. Phys. 10, 2156 (1969); and The Calculation of the Irreducible Characters of the Symmetric Group in Terms of the Compound Characters, (Published as Blaha's Lemma in D. E. Knuth's book: *The Art of Computer Programming Vols. 1 – 4*).

In the early 1980's Blaha was also a pioneer in the development of UNIX for financial, scientific and Internet applications: benchmarked UNIX versions showing that block size was critical for UNIX performance, developing financial modeling software, starting database benchmarking comparison studies, developing Internet-like UNIX networking (1982) and developing a hybrid shell programming technique (1982) that was a precursor to the PERL programming language. He was also the manager of the AT&T ten-year future products development database. His work helped lead to commercial UNIX on computers such as Sun Micros, IBM AIX minis, and Apple computers.

In the 1980's he pioneered the development of PC Desktop Publishing on laser printers and was nominated for three "Awards for Technical Excellence" in 1987 by PC Magazine for PC software products that he designed and developed.

Recently he has developed a theory of Megaverses – actual universes of which our universe is one – with quantum particle-like properties based on the Wheeler-DeWitt equation of Quantum Gravity. He has developed a theory of a baryonic force, which had been conjectured many years ago, and estimated the strength of the force based on discrepancies in measurements of the gravitational constant G. This force, operative in D-dimensional space, can be used to escape from our universe in "uniships" which are the equivalent of the faster-than-light starships proposed in the author's earlier books. Thus travel to other universes, as well as to other stars is possible.

Blaha also considered the complexified Wheeler-DeWitt equation and showed that its limitation to real-valued coordinates and metrics generated a Cosmological Constant in the Einstein equations.

The author has also recently written a series of books on the serious problems of the United States and their solution as well as a book on the decline of Mankind that will follow from current social and genetic trends in Mankind.

In the past twenty years Dr. Blaha has written over 80 books on a wide range of topics. Some recent major works are: *From Asynchronous Logic to The Standard Model to Superflight to the Stars*, *All the Universe!*, *SuperCivilizations: Civilizations as Superorganisms, America's Future: an Islamic Surge, ISIS, al Qaeda, World Epidemics, Ukraine, Russia-China Pact, US Leadership Crisis, The Rises and Falls of Man – Destiny – 3000 AD: New Support for a Superorganism MACRO-THEORY of CIVILIZATIONS From CURRENT WORLD TRENDS and NEW Peruvian, Pre-Mayan, Mayan, Anatolian, and Early Egyptian Data, with a Projection to 3000 AD,* and *Mankind in Decline: Genetic Disasters, Human-Animal Hybrids, Overpopulation, Pollution, Global Warming, Food and Water Shortages, Desertification, Poverty, Rising Violence, Genocide, Epidemics, Wars, Leadership Failure.*

He has taught approximately 4,000 students in undergraduate, graduate, and postgraduate corporate education courses primarily in major universities, and large companies and government agencies.

He developed a quantum theory, The Unified SuperStandard Theory (UST), which describes elementary particles in detail without the difficulties of conventional quantum field theory. He found that the internal symmetries of this theory could be exactly derived from an octonion theory called QUeST. He further found that another octonion theory (UTMOST) describes the Megaverse. It can hold QUeST universes such as our own universe. It has an internal symmetry structure which is a superset of the QUeST internal symmetries.

Recently he developed Octonion Cosmology. He replaced it with HyperCosmos theory, which has significantly better features. He developed a fractionalization process for dimensions, particles and symmetry groups. He also described transformation that reduced particles and dimensions to a far more compact form. He also developed a precursor theory ProtoCosmos that leads to the HyperCosmos.

The author showed that space-time and Internal Symmetries can be unified in any of the ten HyperCosmos spaces in their associated HyperUnification spaces. The combined set of HyperUnification spaces enable all HyperCosmos dimensions to be obtained by a General Relativistic transformation from one primordial dimension in the 42 space-time dimension unified HyperUnification space.

At present the author devel;oped the Cosmos Theory that incorporates ProtoCosmos Theory, HyperCosmos Theory, Limos Theory, Second Kind HyperCosmos Theory and HyperUnification Spaces. He has introduced PseudoFermion wave functions and theory, He has related Cosmos Theory to Regge trajectories of spaces, parton theory, Veneziano amplitudes, Fibonacci numbers and Ramsey numbers. He has calculated an approximation to the difficult R(n,n) Ramsey numbers.

He has developed a Gambol Model that successfully accounts for e-p deep inelastic scattering, fundamental particle resonances, hadron scattering, and the inner structure of particles based on confinement through Casimir forces of ideal gambol gases. The Gambol Planckian Distribution was derived.

He has applied the Gambol Model to particles, universes, and the Cosmos of universes. He showed that the Cosmos may have a distribution of 23 universes corresponding to various Cosmos spaces.

Recently he showed that Cosmos Theory follows from the number of independent asymmetric tensors in a dimension r. He also showed the close parallel between the form of γ-matrices and Cosmos Theory dimension arrays. The closeness suggested that dimension arrays have the same importance as γ-matrices for fermions.

He demonstrated that the pressure of fermions within a space of dimension r balances the Casimir vacuum energy force for 18 dimensions. He showed that $2e\pi$ = 17.02 marks the critical point where pressure balances Casimir force, which implies r = 18 is the highest dimension Physical Cosmos space. The dimension $2e\pi$ appears to set the approximate dimension for Cosmos spaces with dimension array size $2^{r+4} \cong (17.02/8)^{r+4}. \cong (e\pi/4)^{r+4}. \cong 2.13^{r+4}$.

Now he has found the sequences of Coupling Constant values and fermion masses in the Standard Model and UST. This book unifies the Cosmos spaces spectrum, the coupling constant spectrum and the fermion mass sequences in a fundamental scaling differential equation.

www.ingramcontent.com/pod-product-compliance
Lightning Source LLC
LaVergne TN
LVHW071339210926
848486LV00003B/193

* 9 7 9 8 9 9 8 6 0 7 1 3 4 *